GERMAN NIGHT FIGHTERS IN WORLD WAR II

Ar 234-Do 217-Do 335-Ta 154-He 219-Ju 88-Ju 388-Bf 110-Me 262 etc.

Manfred Griehl

Left page:
A Ju 88 R-1 with BMW 801 powerplants. It is in a museum at Hendon, England today.

Translator's note: The term "crazy music" in English only approximates the German meaning, as the word "schräg" properly means "oblique", referring to the angle at which the guns were mounted, and is also a slang term equivalent to the use of the word "crazy" in the early days of rock music.

SCHIFFER MILITARY HISTORY
Atglen, Pennsylvania

PHOTO CREDITS

Anders Collection
Kruse Collection
Dabrowski Collection
Air Archives
Doernier Factory Archives
MBB Factory Archives
Dressel Collection
Lutz Collection
Francella Collection

Van Mol Collection
Heck Collection
Nowarra Collection
Jayne Collection
Petrick Collection
Junkers Factory Archives
Author's Collection
Kössler Collection

Special thanks for their support to Mr. Preuschen, Mr. Sander, the colleagues of the German Museum in Munich, the Dornier GmbH, as well as the Research Group for the History of Air Travel.

Originally published under the title "Deutsche Nachtjager im Zweiten Weltkrieg", copyright Podzun-Pallas-Verlag, 6360 Friedberg 3 (Dorheim), © 1986, ISBN: 3-7909-0294-2.

Translated from the German by Dr. Edward Force.

Copyright © 1990 by Schiffer Publishing.
Library of Congress Control Number: 89-063354.

All rights reserved. No part of this work may be reproduced or used in any form or by any means—graphic, electronic, or mechanical, including photocopying or information storage and retrieval systems—without written permission from the publisher.

The scanning, uploading and distribution of this book or any part thereof via the Internet or via any other means without the permission of the publisher is illegal and punishable by law. Please purchase only authorized editions and do not participate in or encourage the electronic piracy of copyrighted materials.

"Schiffer," "Schiffer Publishing Ltd. & Design," and the "Design of pen and inkwell" are registered trademarks of Schiffer Publishing Ltd.

ISBN: 978-0-88740-200-5
Printed in China

Schiffer Books are available at special discounts for bulk purchases for sales promotions or premiums. Special editions, including personalized covers, corporate imprints, and excerpts can be created in large quantities for special needs. For more information contact the publisher:

Published by Schiffer Publishing Ltd.
4880 Lower Valley Road
Atglen, PA 19310
Phone: (610) 593-1777; Fax: (610) 593-2002
E-mail: Info@schifferbooks.com

For the largest selection of fine reference books on this and related subjects, please visit our web site at: www.schifferbooks.com. We are always looking for people to write books on new and related subjects. If you have an idea for a book please contact us at the above address
This book may be purchased from the publisher.
Include $5.00 for shipping.
Please try your bookstore first.
You may write for a free catalog.

In Europe, Schiffer books are distributed by
Bushwood Books
6 Marksbury Ave.
Kew Gardens
Surrey TW9 4JF England
Phone: 44 (0) 20 8392 8585;
Fax: 44 (0) 20 8392 9876
E-mail: info@bushwoodbooks.co.uk
Website: www.bushwoodbooks.co.uk

This Bf 110 G-4 night fighter lies in the woods near Unterhaching and shows damage to its radiator. Note the two MK 108 as bow armament and the antenna dipoles painted in warning colors.

German Night Fighters 1939-1945

Just a few months after the establishment of the first night fighter units, they were changed into day fighter units. Only the growing numbers of night attacks caused the hasty establishment of several units with twin-engine night fighters in the spring of 1940. At first these were organized as "light night fighters", i.e., with searchlight support. As of 1941 the "dark night fighters" took on greater importance through their use of new types of electronic range-finding and navigational equipment. Air defense districts, known as "sky beds", and searchlight zones were set up purposefully. At the same time, offensive long-distance night attacks over England were begun.

Not only did the number of Royal Air Force (RAF) raids increase steadily in 1942, but the first four-engine bombers also appeared in the night sky. Large-scale attacks, such as that on Lübeck on March 29, 1942, made the power of the RAF tangible. The Luftwaffe replied with improved "Lichtenstein" target-seeking devices and the new-type "crazy music" armaments.

After the first 1000-bomber attack on Cologne on May 30/31, 1942, the dark night fighters gained an even higher priority. But the air offensive against the Ruhr region and the English squadrons that formed streams of bombers continued to show their overpowering effect.

A signal was given by the large-scale attacks on Hamburg in July of 1943. Strips of aluminum, known in German as "Düppel", blinded the German radar devices. Incendiary bombs and firestorms spread unrest and terror. The troops called for improved aircraft and non-disturbable electronic devices.

But waiting for large-series production of the He 219 and Ta 154 seemed hopeless. Until the war's end, almost all units had to go on using the often-improved Bf 110 and Ju 88, which in time received the new SN-2 and Naxos devices.

In the late summer of 1943 the successes of Germany's air defense increased again. The reasons for it were not only the improved radar systems within the parameters of the "tame pig" process, the "sky bed" process, but also the single-engine night fighters, the so-called "wild pig". In addition, the broad-area night pursuit fighting was strengthened, and for a short time long-range night fighting over England was restored. At the end of 1943 there were five fighter divisions in the West and the German territory, with a total of 1200 day and night fighters, which operated from a number of scattered air bases.

Between November 18, 1943 and March 24, 1944, the air war raged over Berlin and gave a foretaste of the coming events. Mosquito bombers flew almost unilluminated, so they could not be intercepted. The promised high-performance night fighters needed to fight them off were still lacking. Meanwhile the year of 1944 was characterized by diminishing success and increasing losses among the night fighter units. Despite the greatest production of night fighters during the war, the training of young pilots could not keep up. Then too, fuel was lacking. The He 219, finally put into series production, came to the night-fighter squadrons too late and in too-small numbers. Planes like the Ta 154, Me 262 and Ar 234-NJ were never produced in quantity.

Thus the burden of air defense in 1945 remained on the Ju 88 and Bf 110.

Ju 88 R-1 of the 10./NJG 3.

Arado Nachtigall

As of July 1944, Arado worked on plans for a two-seater Ar 234 night fighter, of which 30 planes were to be produced at first.

The first model with Weapons Holder (WB) 151 and Radar Device (FuG) 218 was the Ar 234 B-2 with factory number 140145. The plane was reequipped in Werneuchen and Oranienburg beginning on September 20, 1944. Despite the prevailing shortage of materials, the electronic equipment could be produced for the first construction series. Yet the Air Ministry (RLM) stopped the reequipping to meet the urgent need for KG 76 jet bombers by their units.

For the time being, there remained the one model plane. A test command was established on November 11, 1944, under the command of Captain Bisping, and rebuilding was ordered to begin again. On February 23, 1945 Captain Bisping and his radar operator crashed their Ar 234 B-2/N on takeoff. Only on March 26, 1945 did Captain Bonow of Command 388 take command of the test unit. A few weeks before the war ended, the Arado command received additional jet planes, but there was no success with them in service.

Along with the "Nachtigall" (nightingale), the Ar 234 C-3 was to be rebuilt as a night fighter in 1945, and the C-7 was likewise to be set up as a high-performance fighter. The Ar 234 P-1 to P-5 versions also remained projects, though First Lieutenant Welter had already spoken in favor of Ar 234 night fighters with two-level cabins.

Above: An Ar 234 B-2 with two Ju 88 G-6 (620313 and 620557) at Manching airfield late in April of 1945.

Right: Drawing of the Ar 234 B-2/N auxiliary night fighter with underslung "Magirus Bomb" (the WB 151 weapons holder) and FuG 218 antennae at the bow.

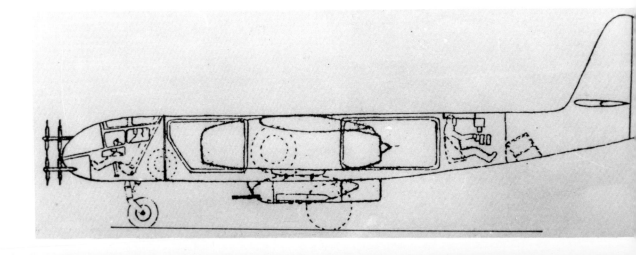

Do 17 and 215 as Night Fighters

The development of the Dornier night fighter took place in three steps:

a) Do 17 Z-7 as a test model,
b) Do 17 Z-10 as a series production model,
c) Do 215 B-5 with DB 601 powerplants.

The first and presumably only Do 17 Z-7 was tested by the I./NJG 2. With it, First Lieutenant Streib was able to shoot down a Whitley bomber on July 20, 1940. Only a small series was built on the basis of the Do 17 Z-10. Its bow armament consisted of one MG 151 and four MG 17. Most Z-10 planes went to the I./NJG 2, where up to seven of these night fighters were in service.

As a test plane, the Z-10 with registration CD+PV was equipped experimentally with the Lichtenstein Radar FuG 212. As better night fighters appeared, the service time of most Do 17 Z-10 planes came to an end.

The next appropriate step was the reequipping of the Do 215 B bomber, which had been developed from the Do 17 Z. Here several planes received the armed nose of the Do 17 Z-10. First Lieutenant Becker scored a night kill during the night of August 8/9, 1941 with a Do 215 B-5 (G9+OM), with the help of FuG 202 B/C equipment. Several Do 215 B-5 planes were still flying in the spring of 1953 with the staff of the II./NJG 1 and the 14./NJG 1. Their registrations were G9+AM and G9+PY.

Some of the Do 215 night fighters were equipped with the "Spanner" type of night vision devices, which were built into the windshields. The suggestion of the Werneuchen Test Center on this matter is of

Below: Under a thick camouflage net is one of the few Do 17 Z-7 night fighters, "Kauz I". The searchlight installed in the bow is easy to see.

interest, as they advised the vertical installation of up to eight MG 151 and a Spanner device for use against bombers.

Left: The Do 17 D-7 model with the I./NJG 2 in the Cologne-Düsseldorf area; the BRAMO powerplants are being warmed up.

Below: A Do 215 B-5 of the 5./NJG 2 painted black for night service. Easy to see are the two 20-mm guns mounted under the fuselage, with muzzle fire dampers.

Below: Three-way drawing of the Dp 215 B-5 "Kauz III" night fighter with Lichtenstein radar system.

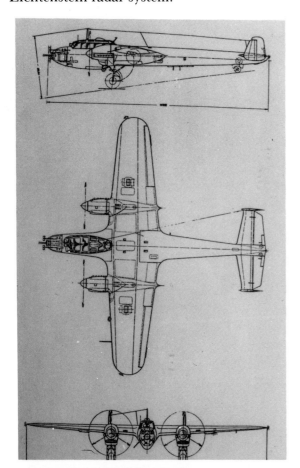

Dornier Do 217 J and N

In October of 1940 it was decided to change Do 217 E planes into night fighters with BMW 801 or DB 605 engines. The planes were designated Do 217 J or Do 217 N.

Production of the J-1 began in March of 1941. The high point was reached in May of 1942, after which production fell astonishingly quickly, since the RLM preferred the Ju 88 C-6. This was one reason why the Do 217 N was tested only as of November 1942. The model planes were Do 217 N-01 to N-03 (GG+YA to GG+YC), which already had, in part, the improved MG 151/20 in place of the four FF machine guns. The first planes without a C gun position were designated Do 217 N-1/U1. In the N-2 the defensive armaments were finally eliminated completely. The planes had also had all other unnecessary equipment removed. Resulting from a suggestion of Captain Schönert, angled weapons were installed in the Do 217 for the first time. In July of 1942, three Do 217 J planes were equipped with four MG 151/20 apiece and tested by Schönert's 3./NJG 3.

The first Do 217 N-2 (factory number 174, PE+AW) already had the angled guns, which had meanwhile been improved. The MG 151/20 was now standard bow armament. By the end of 1943, 157 Do 217 night fighters, mainly of the J-1 and J-2 types (with Lichtenstein radar), were produced. As of the spring of 1943, the version with DB 603 engines slowly came to prevail. After some 340 Do 217 J and N planes were made, production ended once and for all.

The first Do 217 J-1, still without radar equipment, were sent to the 4./NJG 1 in 1942 as overweight night fighters.

Above: Do 217 J-2 at an intermediate stop in Munich-Riem during the winter of 1943/1944.

Left: Side view of the first model of the Do 217 J-2, a rebuilt Do 217 E-1 (factory number 1051), with FuG 202 system.

Two Do 217 J planes, J-1 above, J-2 below, taking off.

Later the 8./NJG 2, the 3./NJG 3 and the 6./NJG 4 received the Dornier planes. Besides frequent powerplant problems, the crews make do with not always convincing flying characteristics. Basically, the Do 217 night fighter remained what it had been, a bomber. The landing gear were weak points for a long time. Problems with engine delivery and technical problems with the DB 603 compelled Dornier to put finished units aside, to remain useless without motors.

In October of 1943 numerous Do 217 night fighters were concentrated in Squadrons 4 and 100 and Training Squadron 101, in order to simplify the supply of spare parts. Some of the planes were turned over to Luftwaffe courier units. Other night fighters, because of the DB 603 powerplant shortage, were finally fitted with the lower-performance BMW 801 motors. One example was the Instructional Group IV./NJG 101 in Hungary. There Captain Krause scored 12 of his 28 air victories with the Do 217 N. But the ranking Do 217 night fighter ace was still Captain Schoenert.

In addition to the two units already named, the Do 217 J and N also saw service with the staff of IV./NJG 2, the 4. and 5./NJG 3, the 11./NJG 4, the staff of NJG 100, the 9., 11., 14., and 18./NJG 101 and the 18./NJG 200.

Above: These Dornier Type J and N night fighters belonged to the IV./NJG 101, a training unit.

Right: Side view of the Do 217 N-2 model without built-in flame-extinguisher pipes.

Above: Inside view of the Do 217 N-1. The removed pilot's seat allows a view of the instrument panel.

A plane in service with Night Fighter Squadron 100 with four MG 151/20 mounted at an angle, flame extinguishers and Lichtenstein radar.

Upper left: Do 217 J-2 with FuG 202 and fire dampers in front of the FF machine guns.

Lower left: A Do 217 J reequipped as a training plane.

Lower right: Zero-series plane with factory camouflage paint and FuG 101 installed in the outer wings.

The High-Performance Do 335 Night Fighter

As of January 1943, the request for the Do 335 as a destroyer was heard. Ten months later the RLM also requested a night-fighter version.

Since the Do 335 was supposed to be in service to fight off the expected invasion, such plans were ignored for the time being. Only on November 15, 1944 was the creation of an initial night-fighter model authorized, while all other possible uses of the Do 335 were cancelled. The night-fighter version was designated Do 335 A-6. The first Do 335 V 10 model could not be rebuilt in time at the Heinkel works in Heidfeld, since the factory was fully occupied with the preparation of the "Volksjäger".

Although the special development commission (ESK) spoke against night and bad-weather action for the Do 335 A-6, the first test plane was finished at the end of January 1945. In spite of that, series production of the night fighter was advocated in 1945, even if only in small numbers. Later a high-performance night fighter, powered by DB 603 E motors, was considered. The Do 335 was also intended for use against Mosquito bombers, since the Ju 88 G-7 was still lacking, there were problems with the Ta 154, and the Me 262 B-1a/U1 did not meet the specifications.

The Do 335 V 15, V 17 and V 22 were also planned as night fighters and were close to completion in April of 1945. Only a test model reached final assembly and was later tested in France.

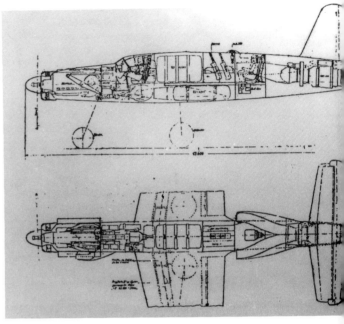

Right: One of the many Dornier night-fighter projects. Notable: angled weapons and HeS 011 jet powerplants.

Below: The rebuilt Do 335 V10 that was tested in France as a two-seat night fighter.

Cockpit arrangement of the Do 335 B-2 destroyer, to which the later night fighter was to correspond.

Upper right: One of the few Do 335 A-11 planes considered for rebuilding into A-6 type auxiliary night fighters.

Lower right: A fuselage of the Do 335 A-11 no longer finished in 1945.

Above: The experimentally reequipped Bf 109 G-6/N with the antennae of the FuG 217 "Neptune" system.

Below: Another model with FuG 350 "Naxos Z" system housed under the glass canopy.

Single-Engine Night Fighters

For short-range night fighting, many Bf 109 G and Fw 190 A planes were reequipped or changed in production as of the summer of 1943. Besides the units of JG 300 to 302, many of the single-engine fighters were used by NJG 10 and 11. Even the heavily armed Bf 109 G-6/R6 (with MG 151 nacelle armament) and the Bf 109 G-14/AS turned up.

Even a few DH Mosquitos could be shot down with the Bf 109 G in 1943/44. Major Müller, whose plane even had an MG 151 mounted at an angle, proved to be especially successful. In 52 flights against the enemy, he scored a total of 23 aerial victories.

Besides the FuG 216V, some Fw 190 A-8/A-9 also had radar devices of the FuG 217J2 and FuG 218J3 types. The Focke-Wulf fighters were usually reequipped and put in service by the test center in Werneuchen. The I./NJG 11 and parts of the JG 300 operated from Bonn-Hangelar with the Fw 190 night fighter. Some of the planes had been equipped with flame extinguishers. Unlike the Bf 109, the Fw 190 scarcely caused problems when used as a night fighter. The Messerschmitt fighters suffered numerous landing-gear failures in night service.

As of the latter half of 1944, single-engine short-range night fighters came to be regarded as only of secondary importance, since their range and length of action could allow only limited success. Thus in the last months of the war many night fighters became night assault planes, a fate that many another NJ crew was not spared either.

Above: First Sergeant Döring's plane of the 2./JG 300 in August of 1943, a Bf 109 G-6 with two nacelle-mounted MG 151.

Right: Main instrument panel of a Bf 109 G-6/N night fighter with integrated sight device in the upper row of instruments.

Above: The "White 9", an Fw 190 A-8 of the 1st Unit of NJG 10. A light shield was mounted over the exhaust port.

Upper left: These Fw 190 A-6 planes of the 2./JG 300 were equipped with FuG 217 J as of 1943.

Lower left: During testing in Werneuchen: a model of the Fw 190 A-8 night-fighter.

Ta 154, the German Mosquito

From the Focke-Wulf design of the Ta 211 high-speed bomber there grew the Ta 154 high-performance fighter, which had its first flight at Hannover-Langenhagen on July 1, 1943. Its flying characteristics were quite encouraging. But the Jumo 213 powerplants wanted by Professor Tank were not ready until the summer of 1944. So Jumo 211 F/N motors had to be used. Only one plane had Jumo 211 R powerplants. On December 22, 1943 Troop Testing Command 154 was established. Because of problems with the wooden parts, the RLM halted series production in the summer of 1944 and cancelled the entire model in September. Only one of eight series planes was officially received by the Luftwaffe. There were also twelve test planes, five incomplete fuselages and several prototypes under construction. Factory numbers 320008 to 320010 were rebuilt into bad-weather fighters (A-2/U4) and four others into Ta 154 A-4 night fighters.

Along with three Ta 154 planes with the III./NJG 3 in Stade and a fourth plane with the staff of the I./NJG 3 in Grove, Ta 154 planes served only with the III. (Completion)/JG 2 in Lechfeld.

One Ta 154 night fighter with the antennae attached in the middle of the wing and angled wing tips crashed at Stade on April 30, 1945.

In 1944 several manned explosive carriers and "mistletoe bottoms" were made of available components in Central Germany during 1944. But these machines did not see active service.

Above: The wooden mockup of the Focke-Wulf night fighter, later the Ta 154, with FuG 202 and four machine guns.

Below: Front view of the Ta 154 V1 at the spacious Hannover-Langenhagen airfield.

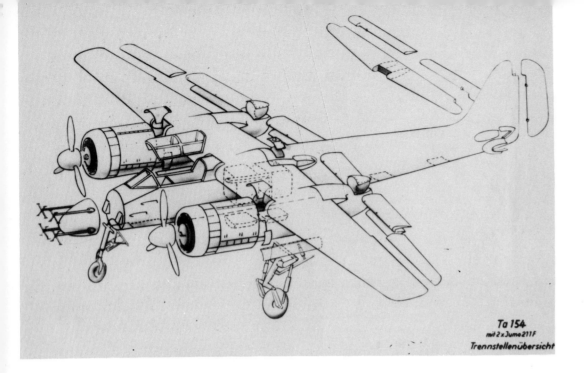

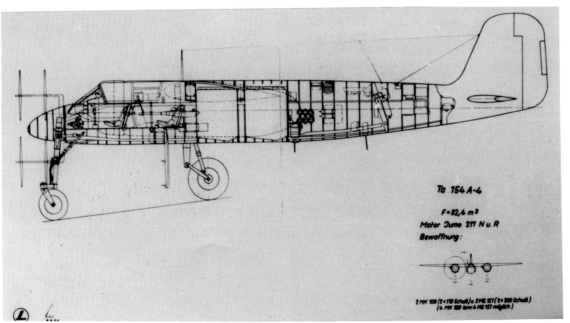

Above: Professor Tank after his flight with the Ta 154 V1 in Hannover on July 7, 1943.

Upper left: Exploded drawing of the Ta 154 V1 to V4 with Jumo 211F motors.

Left: Side view of the Ta 154 A-4, which was produced in small numbers as of the end of 1944.

Upper left: Side view of the Ta 154 A-O (factory number 120005), which was later rebuilt into an A-4 type.

Upper right: The third Ta 154 (TE+FG) with FuG 212 C-1, destroyed in an air raid on August 5, 1944.

Lower left: At Lechfeld at the end of April 1945 this Ta 154 was photographed; it had been used to train jet pilots.

Focke-Wulf Night Fighter Projects

On to the future... numerous Focke-Wulf projects seem to have been born under this sign. But this was particularly true of the development of modern high-performance night fighters.

In November of 1944 the brief description of a "night fighter and bad-weather destroyer with Jumo 222E/F and two BMW 003 turbines" was presented. The possibility of turning off the two jet engines and flying with a piston engine provided a great range and over seven hours' flying time. The top speed was 880 kph.

In January of 1945 three further projects were introduced: a night fighter with DB 603 N, Jumo 222 C/D or As 413 powerplants and methanol-water (MW) 50 system. For fast flying, one used two additional turbines under the wings. Interchangeable cabin armament was planned, with MK 112, 103, 108 or 213 guns. The radar equipment included, among others, the modern Berlin target-seeking device.

The last project, the night fighter with two HeS 011 powerplants, came out at the end of January 1945. It was a cantilevered low-wing design with jet turbines built into the bow. The machine, weighing 11,000 kg, was to be armed with four MK 108 and use the FuG 244 system. Its calculated top speed was over 930 kph. None of these projects ever progressed beyond the project stage.

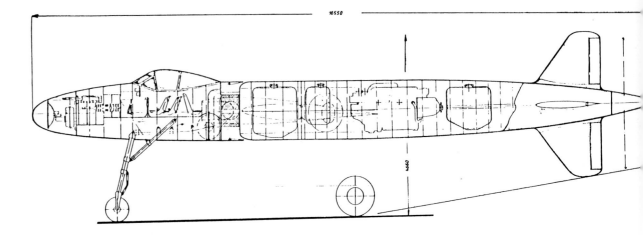

Above: Project drawing of a heavy night fighter with three-man crew and 21-meter wingspan.

Below: Side view of the night fighter with pressurized cabin, angled guns and two Heinkel He 011 jet turbines.

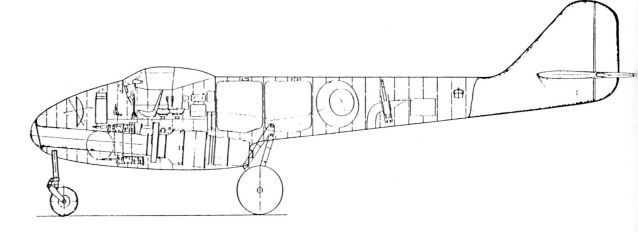

Heinkel Uhu

In July of 1941 the Heinkel projects P 1055/1056 gave birth to the He 219 night fighter. The first flight of He 219 V1 took place on November 6, 1942. The results of the early test flights were satisfactory. The favorable attitude of General Kammhuber led to a contract for large-series production.

But on September 17, 1943 Milch also spoke in favor of hastening the completion of a high-performance night fighter, although many factors favored the production of the Ju 88 C, which was easier to build. In the end a compromise was made which provided for series production of both models. Kammhuber's and Galland's encouragement of the much more progressive He 219 were largely in vain. Then too, fitting stronger powerplants (Jumo 213 or 222) could not be done at the time. The same problem plagued the constructors as to the request for fuselage armament of MK 103 and MK 108 guns. So for a long time there was only step-by-step reequipping of the available zero-series models. The He 219 A-O planes, powered by two DB 603A motors, were delivered with various types of armament (R), such as two MK 108 (R1) or four of the same (R2). The A-O series also differed in their use of FuG 212 C-1 and C-2 radar equipment. The A-o/R6 version, model for the He 219 A-5 series, had not only improved radar equipment but also two MK 108 mounted at an angle. The He 219 A-1 with DB 603E powerplants was a bomber project and, like the A-3 with DB 603G engines, had no more chance of materializing than the A-4 with Jumo 222, a

In the summer of 1944 at Munich-Riem airfield. Before the complex of buildings stands this He 219 A-5/R1.

The DV+DL, one of the He 219 A-5 planes; it originated from the He 219 A-O/R6 and was equipped with FuG 220 (SN 2) radar.

faster reconnaissance plane. What was produced in series was the He 219 A-1. These planes had an armament of two MK 103 in the fuselage and two angled MK 108 as "crazy music". The SN-2 radar apparatus was installed.

The A-5/R1 to A-5/R4 versions were also built in limited numbers—with DB 603A and E powerplants. All these planes used the FuG 220 and angled MK 108. They differed only in their use of two MK 103, MK 108 or MG 151/20 in the weapon housing.

A considerably lightened He 219 A-6 with four MG 151/20 was intended to see service as a Mosquito chaser. By the end of the war, the Mosquito hunter had finally been completed, with Jumo 213 E motors and auxiliary injection.

The last series version that was still produced in small numbers was the He 219 A-7 with DB 603 G motors. Until the A-7/R2 and Re versions, these had no angled guns. In 1945 only a model of the He 219 A-7/R6, with Jumo 222 powerplants, could be built.

After several test models, the NJG 1 received most of the 195 UHU's delivered to the Luftwaffe. Another 73 planes served experimental purposes or were destroyed in air raids. But the results achieved by July 17, 1944 spoke well for the He 219. They had shot down seven Mosquitos and 104 other enemy planes. But the Heinkel night fighter never took part in the defense of German air space (RLV) in large numbers.

Below: the ejection-seat test plane DV+DI, with its partly removed cabin canopy.

Right: Drawing of the Mosquito chaser He 219 A-6, with only four machine guns, which was produced in small numbers.

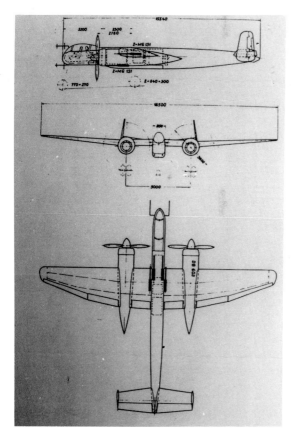

23

Junkers Night Fighter

After the first studies that considered rebuilding the Ju 88 A-1 bomber into a destroyer with Jumo 211 B motors, the C-1 to C-5 versions were soon suggested. After the Ju 88 C-1, produced only in individual models, came the C-2 with its five-barreled armament arrangement. The C-4 corresponded to the C-2 but was powered by two Jumo 211 F motors. The C-5 flew as a destroyer with BMW 801 powerplants. The widespread Ju 88 C-6 was the variation of the Ju 88 A-5 that had first been planned with a weapon housing that could be jettisoned instead of the cabin armament. The first night fighters were able to fly in all weather, but did not yet have target-searching radar. But their great range allowed them to make long night flights over England and North Africa. The first model of a Ju 88 C with Lichtenstein radar was contracted for in March of 1943 and reequipped shortly afterward. At the same time, a much-needed hastening of Ju 88 C-6 production was planned, since the Bf 110 night fighter did not live up to specifications and was not able to meet the need. General Kammhuber considered the Ju 88 C-6 to be outmoded and quickly urged that production be concentrated on the Ju 88 G-1. Because of the shortage of BMW powerplants, it took months before series production with the desired motors could take place. As of September 1943, the Ju 88 C night fighters could finally be equipped with SN-2 radar, and the call for further increases in performance was heard again.

Left: One of the early attempts to install additional fixed weapons (TO of the KG 51, Dr. Stahl).

Below: Three-way drawing of the early Ju 88 C-2 night fighter with Jumo 211B powerplants.

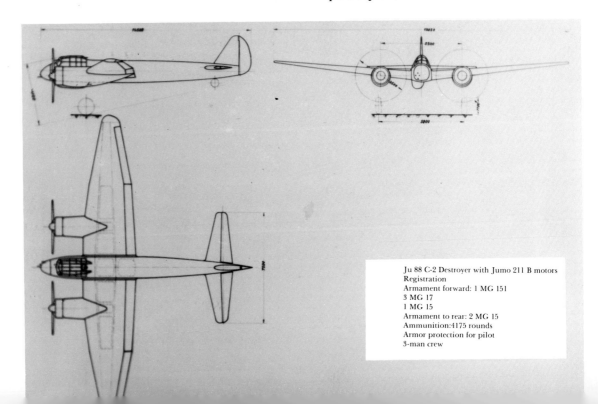

Ju 88 C-2 Destroyer with Jumo 211 B motors
Registration
Armament forward: 1 MG 151
3 MG 17
1 MG 15
Armament to rear: 2 MG 15
Ammunition: 4175 rounds
Armor protection for pilot
3-man crew

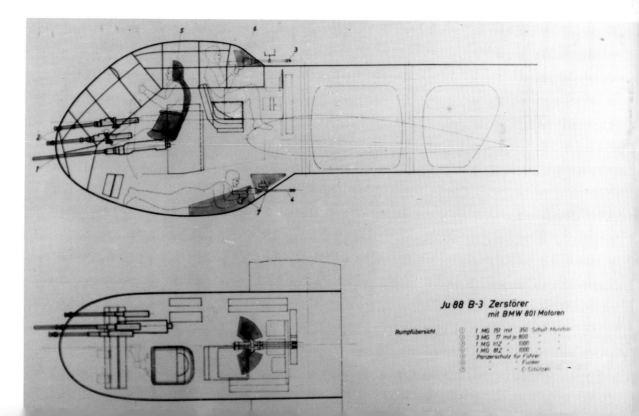

Above: Bow of model plane Ju 88 V19, also called Z19 (factory number 0373), after an additional 20-mm gun had been installed.

Upper right: Factory test plane Ju 88 C-6 (DU+GO), with which the angled armament of two MG 151/20 was tested in particular.

Lower right: The Ju 88 B-3 destroyer and night fighter with BMw 801 motors did not go into production.

The suggestion made by Field Marshal Milch a year before of equipping the Ju 88 C-6 with BMW 801 and the GM 1 system failed primarily for lack of powerplants. Only in March of 1943—after the building of a few models—did the Ju 88 R series go into production.

Also tested was a Ju 88 C-6 with two Jumo 213 powerplants. By this means the speed was increased by almost 90 kph. Since these powerplants were available only in insufficient numbers, a refitting had to be postponed. Similar problems kept the plans for a Ju 88 C-6 with TK 11 supercharger, a high-altitude fighter, from progressing beyond the project stage.

At the end of 1943, production of the Ju 88 R-1 and R-2 began. Both series had BMW 801 powerplants and differed only in their radar equipment and engine details. The Ju 88 R-2 was produced with SN 2 radar. Individual planes had angled guns like those of the Ju 88 C, consisting of two MG 151/20. In August of 1944 only the Ju 88 C-6 and R-2 were produced in addition to the Ju 88 G.

The C- and R-series night fighters saw service with the units of almost all night fighter squadrons.

Both series were finally replaced more and more by the Ju 88 G-1 and G-6. Planes that were still available were gradually turned over to the night fighter training units.

Above: One of the Ju 88 R-2 planes with FuG 220 used by the I./NJG 2. Like the Ju 88 C-6, this version still had cabin armament.

Below: Captain Tober's crash-landed plane, a Ju 88 C-6 with "crazy music". The two Jumo 211 F motors broke loose from their mountings.

Above: A Ju 88 C-6 equipped with FuG 212 BC and FuG 220, but still having day sight protection.

Upper right: A Ju 88 C-4 of NJG 2 in Sicily.

Lower right: This Ju 88 C-6 with Lichtenstein BC 2 (FuG 202) radar, camouflage paint and flame-extinguisher pipes belonged to NJG 100.

The Ju 88 Gustav

The production of 100 Ju 88 G night fighters was requested by completion planning as early as February 1942. In May of 1943, 700 Ju 88 G planes were ordered, in 1944 more than 1800.

On October 26, 1943 it was clear beyond doubt that the Ju 88 G, like the Ju 88 R-2, could not be built on schedule or delivered in the specified quantities because of the lack of the right powerplants.

The first G series built had the front fuselage of the Ju 88 A-4, the main fuselage of the Ju 188 E-1, the wings of the Ju 88 D-1 with BMW 801 motors, and at first a fixed armament of six MG 151/20. Since the two guns installed in the bow damaged the antennae when they were fired, they were later eliminated.

In one of the first air battles between Ju 88 G-1 and Mosquitos on December 3, 1943, the results were more or less equal: three Ju 88 G and two DH mosquitos were lost!

The Ju 88 G-1 was also built individually with GM-1 (auxiliary injection system) and delivered in small numbers as type R-1.

The subsequent Ju 88 G-2 consisted of production lots of the Ju 88 G-1, Ju 188 A-2/G-2 and Ju 388. Except for its wings and Jumo 213 powerplants, it was identical to the Ju 88 G-1. A temporary solution until the Jumo 213 E was delivered was represented by the Ju 88 G-3. The night fighter had the wings of the Ju 88 A-4 and was powered by two DB 603 E motors. Only one model plane, the Ju 88 V 105, factory number 710523, was built. The Ju 88 G-4 ranked as a heavy fighter with Jumo 213 A motors and did not go into series production.

The Ju 88 G-5 was cancelled as early as March 2, 1944, since it was thought the Ju 88 G-2 would be available soon.

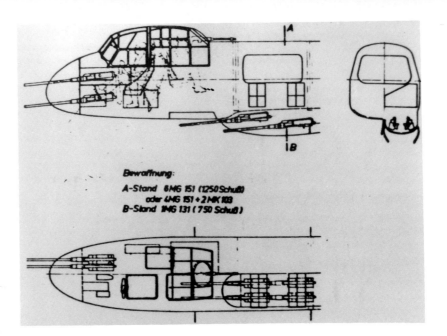

Drawing of the planned series production Ju 88 G-1 with six MG 151/20.

Below: The first model of the Ju 88 G-1 version (factory number 700001), which first flew on June 24, 1943 as V58.

Upper left: This Ju 88 G-1 belonged to an IV. group.

Upper right: A Ju 88 G-1 at the Werneuchen test center. The FuG 227 "Flensburg" and the "Neptune II R" rear warning device are built into the wings.

Lower right: Ferrying by day: a Ju 88 G-1 with FuG 220 flies over Germany.

Other than the Ju 88 G-1, only the G-6 version was produced in large series. The planes were equipped with four MG 151 in the weapon housing and two more as "crazy music" as standard equipment. Except for the powerplants, the plane was almost identical to the Ju 88 G-1. The FuG 220 was usually installed. At the end of 1944 numerous individual models were built, such as the Ju 88 G-6 with Morgenstern antenna and Berlin device. The G-7 version was based on the series Ju 88 G-6. This night fighter had the wings of the Ju 188 and Jumo 213 E powerplants with auxiliary injection (MW 50) and four-bladed airscrews. As well as the high-performance G-7 night fighter, a special Mosquito chaser was to be built in small series. Two initial models were ordered on November 6, 1944. Before their first flight could take place, they were lost in a night air raid on Dessau (March 1, 1945). Two more were still being worked on in April of that year.

The last series model of the Ju 88 G was the Ju 88 G-10, built in Meersburg in the spring of 1945. Though intended to be a long-range night fighter, the planes, of which more than twelve can be documented, were usually used as the bottoms of "mistletoe" combinations at the war's end.

Above: Several Ju 88 G-6 of the IV./NJG 2 are off the beaten track at this airfield, to avoid Allied air raids.

Below: The war is over for NJG 3. The camouflage has already been removed.

Below: This Ju 88 G-6 of the 7./NJG 100 seems to be armed with only one angled MG 151. Note the fire extinguisher near the MG 131 in B position.

Above: Brake failure probably made this Ju 88 G-1 (CI+LN) do a headstand.

Above: A long row of Ju 88 G-6 that fell into Allied hands in 1945.

Upper left: Ju 88 G-6 with SN 2 system and two MG 151/20 as angled armament in North Germany, 1945.

Lower left: Another picture of this plane shows the angled antenna dipoles. Fire extinguishers are missing.

Above: The lack of fuel compelled Ju 88 night fighters to be moved by ox teams in the winter of 1944/45.

Below: The exact use of this Ju 88 G-6 (D5+AZ, 15./NJG 3), with windows in its fuselage, has not yet been made clear.

Above: Despite fire damage, this Ju 88 G-6 shows the FuG 218 V2R "Neptune" attachments and interesting details of the Jumo 213 A-1.

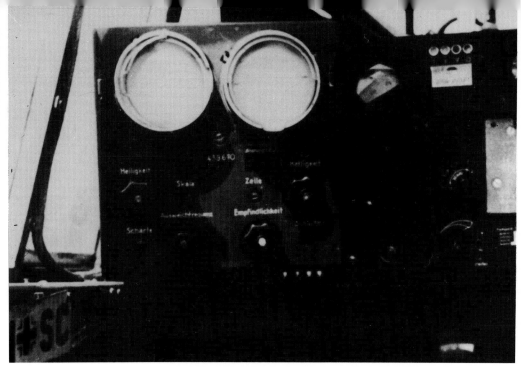

Above:
One of the few Ju 88 G-6 with the modern FuG 240/1 radar apparatus under the temporary covering.

Upper right:
The standard sight device for the FuG 212 and FuG 220 radar system.

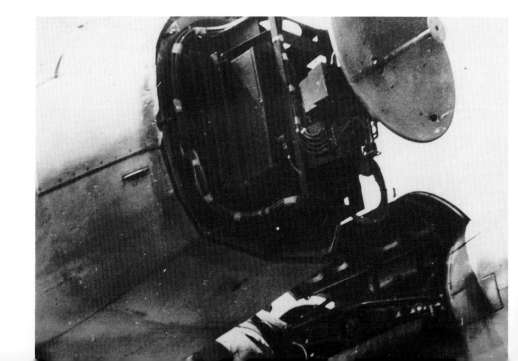

Lower right:
A close-up look at the Berlin device in 3C+MN of NJG 4. Barely ten of these devices were in use.

Above: Numerous Ju 88 G-6 in various camouflage colors, along with He 219 and Bf 110G, at Grove.

Upper left: English soldiers found 3C+PN near Hamburg with its Jumo 213 engines removed.

Left: Six Ju 88 G-6 with SN 2 radar apparatus at Wunstorf in the spring of 1945.

Ju 188 and Ju 388 Night Fighters

In the summer of 1943, Field Marshal Milch often advocated the use of the Ju 188 as a night fighter, since the model "had proved itself as a bomber".

At times General Kammhuber also praised the Ju 188 G-1, production of which was to begin in the fall of 1943. The planes were first planned as destroyers and long-range night fighters. The service model was designated Ju 188 R. Production of models did not take place for a long time. The argument as to whether the He 219 or Ju 188 should be built dragged on from month to month. In the process, the Ju 388 came more and more into the picture, being regarded as a future heavy night fighter. Though the Ju 188 was easier to build, a glazed fuselage nose was not regarded as exactly ideal for a night fighter. The actual battle distance was still not great, despite new machine guns, so that the danger of being hit by flying parts of the enemy plane had to be taken seriously. Then too, the closeness of antennae and bow weapons to each other drew criticism.

So it was the Ju 388 J that continued into 1944 to gain more and more emphasis as the new Junkers night fighter.

As for performance, the Ju 388 surely would have been capable of giving effective service as a night fighter, but the intricate machines could not be produced in sufficient quantities at the end of 1944. Then too, the Ju 388 J was too slow and ponderous to pursue Mosquitos. For the four-engine bombers normally used by the RAF, though, the pressurized cabin and BMW 810 TJ high-altitude powerplants were not necessary. So there appeared only two finished test models, Ju 388 V1 (500001, PE+IA) and V2 (500002,

Model installation of five MG 151/20 and one FuG 212 system in a Ju 188, which also had considerably strengthened cabin armor.

Left and right above: Model construction in the form of a wooden mockup: Ju 388 with FuG 212 and two 37-mm guns.

Left: Cockpit mockup of a Ju 388 night fighter with FuG 212 and movable defensive weapon in the rear of the cabin.

PE+IB). The first prototype flew in May of 1944 in Dessau and was later tested in Tarnowitz with SN 2 radar and MG 151/20 armament. As of January 13, 1945 the plane was in service with Command 388. The Ju 388 V2 had a fuselage housing with two MK 108 and two MG 151/20, two more MG 151 mounted at an angle, and the SN 2 radar device too. As of December 1944 the Luftwaffe took over the testing of this model. The remote-control rear guns originally intended for most Ju 388 versions were not yet fully developed technically and were therefore replaced by dummies in the night fighters. Series production of the night fighter was to begin in January of 1945. A few planes were under construction that spring at the Junkers factory in Leipzig-Mockau. But a heavy air raid made the scheduled production impossible. Three basic versions of the Ju 388 J were planned: the J-1, with two MK 103 and two MG 151, as a daytime destroyer, and with two MK 108 and two MG 151 as a night fighter. Among others, BMW 801 G motors were considered as powerplants.

The J-2 corresponded extensively to the J-1 but was planned with Jumo 222 A/B powerplants. A high-altitude fighter version with Jumo 222 E/F motors also took shape on the drawing board. The Ju 388 J-3 was a high-altitude fighter with Jumo 213 E powerplants.

Data sheet of the Ju 388 J-3 with SN 2 "Morning Star" antenna and Jumo 213E powerplants.

Rear armament mockup with FHL 131 Z with aiming controls and fast fuel dumping system.

Above: This picture of the Ju 388 V2 (factory number 500002) was taken during testing at Rechlin at the end of 1944.

Upper right: Night-fighter mockup of the Ku 388 J with SN 2 radar and FuG 212 radar.

Below: One of the few fuselage bows that were actually built for the Ju 388 J-1.

Messerschmitt Night Fighter

At the beginning of the war, the destroyer was a completely new type of weapon with offensive capability; its main uses were escorting bombers and supporting ground forces. On account of their tactical inferiority to single-engine enemy fighters in aerial war, the Bf 110 was used more and more as a night fighter during the course of the war. As of 1940, the Bf 110 C-1 was delivered with DB 601 A-1 motors. The C-2 production run carried two MG 151/20 at the bow instead of FF machine guns. Its first successes against four-engine bombers were gained by the Bf 110 night fighters on April 10, 1941 (Short Stirling) and June 24, 1941 (Halifax). But their performance advantage over their opponents was slight.

In February of 1941 Bf 110 production hit a high point for the time being; then it fell to nothing by January of 1942, since officialdom was sure the Me 210 night fighter would be available. In the summer of 1942, lost planes could be replaced only in the direst need. Improved types such as the Bf 110 F-2/F-3 with DB 601 F powerplants, some with three-man crews, with angled guns and enlarged rudders, finally reached the troops.

In 1943 the Bf 110 was regarded only as an auxiliary plane. Every week the higher-performance He 219 was expected. New Bf 110 night fighters were meanwhile ready for production, such as the G-2 version with two MG 151 and four MG 17 as bow armament. Either two MG 151 or one BK 3.7 could be fitted under the fuselage.

Right: A sergeant of NJG 1 stands before his night-black Bf 110 C with fourteen shoot-down markings.

Below: Close-up shot of the Bf 110 G-4 (2Z+OP) of the 6./NJG 6, with black auxiliary tanks.

At the same time, substitution of the MK 108 cannon for the too-light MG 17 progressed slowly. A dummy construction with the new armament arrangement was built by the Gotha Wagon Factory. The first model planes went into service with the IV./NJG 1 between April 2 and May 27, 1944, and scored 32 aerial victories there. After the Bf 110 G-3 reconnaissance plane came the Bf 110 G-4, a night fighter with DB 603 motors and FuG 202 radar. Its armament was generally identical to that of the G-2 version. As a rule, the planes were armed with four MG 151 instead of the two MK 108. There were often two MG 151/20 angled guns too. There was then not enough space left for the MG 81Z rear defensive weapon.

The limits of performance capability finally seemed to have been reached with the Bf 110 G-4/U7 night fighter. This plane had SN 2 radar, an auxiliary injection system (GM 1), and a fuselage lengthened by 60 cm. Testing took place with NJG 1.

The angled MG 151 were often replaced by MG FF cannons, which were available in sufficient numbers. In 1943 practical tests were conducted with numerous 30-mm weapons, the MK 101 and a bow armament consisting of two MK 108 and one MK 103. A model plane (factory number 740002) flew between August and December of 1943. Most Bf 110 C saw service with NJG 1 and 3, types F and G with Squadrons 2, 3, 5 and 6. The Bf 110 G was also used by Air Observation Units 1 to 7 and the Norway/Finland Night Fighter Unit.

Above: Radioman's (and gunner's) seat of a Bf 110 night fighter.

Another photo of 2Z+OP.

Above: Another Bf 110 with FuG 212/220 and two 300-liter auxiliary fuel tanks, plus the usual flame extinguishers.

A Bf 110 G-4 in the winter of 1943, with the same equipment, including FuG 212 and 220.

Left: Preparing for takeoff to take on an Allied air raid.

The 8./NJG 3 flew Bf 110 G-4 planes with Lichtenstein radar.

This Bf 110 G-4, with factory number 110087, belonged to the night fighter unit in Finland.

A burned-out Bf 110 G-4 in 1945.

A factory-fresh Bf 110 G-4 fell into the hands of English troops at Braunschweig in 1945 (factory number 160790).

Above: Well concealed from sight from the air: a Bf 110 G-4 with SN 2 radar equipment and auxiliary fuel tanks that can be jettisoned.

Above: A Bf 110 camouflaged with branches.

Lower right: A makeshift trailer for transporting airplane fuselages. Its load: a Bf 110 G-4 "with antlers".

Above: Bf 110 G-4 night fighter with the M1 weapons set consisting of two MG 151/20, plus Lichtenstein radar.

Above: Front view of the same machine in service.

Left: This damaged Bf 110 G-4 belonged to NJG 4, and was to be taken away by rail.

A heavy night fighter of the Bf 110 G-4 type, with two MG 151/20 instead of two MK 108 at the bow. Under the fuselage there is also a 37-mm cannon.

These Bf 110 G-4 planes of Night Fighter Squadron 9, to whose first unit factory number 160128 belonged, were found near Fritzlar at the end of the war.

Me 210 and Me 410 as Night Fighters

Neither the Me 210 nor the improved Me 410 could fulfill the great hopes set on them. Since production of the Me 210 did not begin as planned, the Bf 110 was compelled to stay in service too long. The expected improved Messerschmitt Me 410 night fighter was expected everywhere during the summer of 1943, but its use was greatly delayed in any case and usually limited to service as a high-speed bomber.

In October of 1943 two He 219, one Ju 88 and three Me 410 planes were flown as long-range fighters in an attack on English bombers. They were not able to shoot down any of them. The use of the Me 410 with Air Observation Units 1 to 7 over all of Germany was more successful. Along with the Me 410, mainly Bf 110 G-4 and Ju 88 G-6 planes were being used as well as Ju 88 C-6 at the end of the war. There were also Ju 88 A-4 planes from the Luftwaffe's fighter squadrons, used as target indicators.

In the spring of 1944 the II./KG 51 prepared for long-range night-fighting duty. In one of its actions the group's commander, Captain Puttkammer, shot down five heavy English bombers over the British Isles with his Me 410.

Additional Me 410 planes—likewise without radar—were used by the I./NJG 5 and III./NJG 1. They were generally Types A-1 and B-2. A few Me 410 with GM-1 systems also flew as Mosquito chasers. Since they did not do very well, the Luftwaffe command decided to use the He 219 A-6, Ju 88 G-7 and Ta 154 instead.

The Me 410 was to serve as the basic version for reequipping as a night fighter, which was urgently needed for service in the west plus as a reconnaissance plane.

Aerial view of an Me 410.

Project drawing of an Me 410 night fighter with four MK 108 and two MG 151 as bow armament plus two angled MK 108.

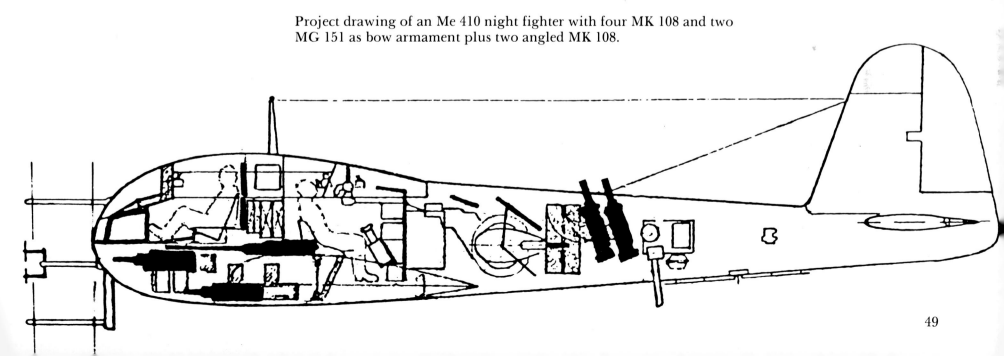

Final Accord: Me 262

A night fighter of the 10./NJG 11, the Welter Command, in May of 1945.

The Me 262 B-1a/U1 and the Ar 234 B-2/N were to fill in as auxiliary night fighters until the Me 262 B-2 and Ar 232 C-7/N arrived. The two-seat Messerschmitt night fighters were rebuilt in small series at the DLH (German Lufthansa) facilities in Berlin-Staaken and turned over to the 10./NJG 11, the former Welter Command. Frequent problems with the fuel tanks made sure that the Me 262 B-1a/U1 generally remained in doubt. First Lieutenant Welter's unit thus made their flights in the greater Berlin area with series Me 262 A-1a. Between December 12, 1944 and May 8, 1945 they shot down some 50 Mosquitos and four-engine night bombers in some 70 attacks on enemy flights.

At the same time, the Upper Bavarian Research Institute in Oberammergau worked on plans for future Messerschmitt night fighters. The project presentation of the Me 262 NJ, with lengthened fuselage, was followed on January 18, 1945 by the project construction description of the Me 262 B-2 with two Jumo 004B-2 jet turbines. This plane, armed with MK 108 guns, was to have a flying time of two and a quarter hours at an altitude of 6000 meters; its performance was between those of the Do 335 A-6 and the Ar 234 C-3/N. On February 12, 1945 there followed the project description of a night fighter with HeS 011 turbines as a variant of the Me 262 B-2a. A little later, on March 17, the design of the 900-kph night fighter with HeS 011 and angled guns was submitted. The series of projects ended on March 27, 1945 with the progressive night fighter powered by two HeS 011 motors. The Me 262 B with prop-jet engines and Bremen-O radar also got no farther than the drawing board.

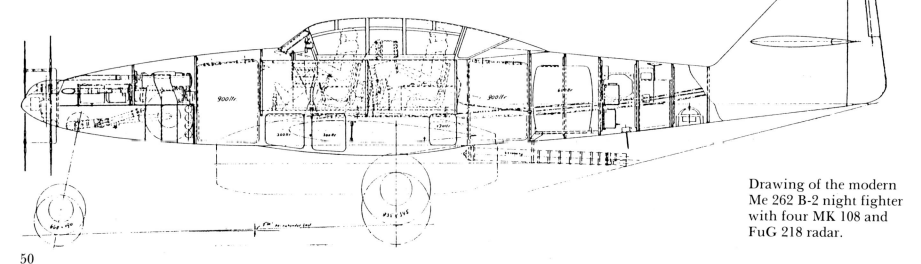

Drawing of the modern Me 262 B-2 night fighter with four MK 108 and FuG 218 radar.

Aibling, May 1945: Ju 88 G-6 and Bf 110 night fighters among single-engine Messerschmitt and Focke-Wulf fighters.

TECHNICAL DATA

Manufacturer	Arado	Dornier	Dornier	Focke-Wulf	Heinkel	Junkers	Junkers	Messerschmitt	Messerschmitt
Type	Ar 234 P-1	Do 217 N-2	Do 335 A-6	Ta 154 A-4	He 219 A-5	Ju 88 G-6	Ju 388 J-1	Bf 110 G-4	Me 262 B-2a
Use	Hochleistungs-NJ	NJ	Hochleistungs-NJ	Hochleistungs-NJ	NJ	NJ	Höhen-NJ	NJ	Hochleistungs-NJ
Crew	2	3	2	2	2	4	3	2 – 3	2
Powerplants	4 x BMW 003 A-1	2 x DB 603 A-1	2 x DB 603 E-1	2 x Jumo 213 E-1	2 x DB 603 E-1	2 x Jumo 213 A-1	2 x BMW 801 J-O	2 x DB 605 B-1	2 x Jumo 004 B-2
Wingspan (meters)	14,40	19,00	13,80	16,00	18,50	20,00	22,00	16,30	12,51
Length (meters)	13,30	18,10	13,85	12,60	15,55	14,95	17,55	12,10	10,65
Height (meters)	4,15	5,03	5,00	3,60	4,40	5,07	4,90	4,00	3,05
Flying weight (kg)	12.050	15.000	10.100	8.450	11.900	14.700	13.760	9.270	7.800
Fuel (liters)	3.700	2.960	2.320	1.500	2.670	3.205	3.280	1.870	3.070
Armament	1 x MG 151 2 x MK 108	4 x MG 151 4 x MG 17	2 x MG 151 1 x MK 103	2 x MG 151	2 x MK 108 4 x MG 151	4 – 6 x MG 151 1 x MG 131	2 x MG 151 2 x MK 108	2 x MG 151 4 x MG 17 1 x MG 81Z	4 x MG 108
Radar device	FuG 240	FuG 202	FuG 218	FuG 220 D	FuG 220	FuG 220	FuG 220	FuG 220	FuG 218
Top speed (kph)	860	515	680	560	570	550	580	500	840
Ascent (meters per second)	14	7	12	10	8	- - -	- - -	8	12

Fw 189 auxiliary night fighter with an angled MG 151.